BENBROOK PUBLIC LIBRARY
3 4267 00004 6834

D1709466

Bridget Heos

New York

To my Mom and Dad, who taught me that no man is an island . . . or even a peninsula

Published in 2010 by The Rosen Publishing Group, Inc.
29 East 21st Street, New York, NY 10010

First Edition

Library of Congress Cataloging-in-Publication Data

Heos, Bridget.
The creation of peninsulas / Bridget Heos.
p. cm.—(Land formation: the shifting, moving, changing earth)
Includes bibliographical references and index.
ISBN-13: 978-1-4358-5301-0 (library binding)
ISBN-13: 978-1-4358-5600-4 (pbk)
ISBN-13: 978-1-4358-5601-1 (6-pack)
1. Peninsulas—Juvenile literature. I. Title.
GB454.P46H36 2010
551.41—dc22

2008051105

Manufactured in Malaysia

On the cover: Peninsulas can be almost any size and form in several different ways.

CONTENTS

INTRODUCTION

Everybody knows that a peninsula is land that is almost entirely surrounded by water but is connected to a larger landmass. They know that Florida is a prime example.

But did you ever wonder how Florida became a peninsula? It's a long story. Millions of years long, in fact. But you probably want the short version.

Here's what happened:

Today, the earth contains seven continents. But 290 to 248 million years ago, the earth formed just one supercontinent, called Pangaea. Of course, nobody was around to call it that. This was even before dinosaurs roamed the earth.

Obviously, the single continent didn't stay that way forever. Continental drift explains why. The continents rest on large tectonic plates. These are sheets of rock 50 to 250 miles (80 to 402 kilometers) thick and ranging in size. (The Arabian Peninsula in Asia rests on its own tectonic plate, for instance, but India shares a plate with Australia.) Forces on the ocean floor and the earth's crust cause the plates to slide along the semi-molten center of the earth, moving the continents.

Florida was formed by marine sedimentation, just one of the many geological processes that create peninsulas.

To fully understand this, imagine you and your friends floating on rafts in a swimming pool. You join your rafts together at the center of the pool. Now, let go and shut your eyes. Maybe somebody dives under your rafts, causing waves. The waves cause some of the rafts to move away from the others. You drift apart. So did Pangaea. Only instead of taking minutes, it took millions of years. And instead of somebody pulling the continents apart, complicated earth processes did the job. Now, say you close your eyes again. This time, if somebody dives under your rafts, the waves might actually unite you and your friends back in the center of the pool. Scientists believe that earth processes will drag the continents back together 250 million years from now, forming Pangaea Ultima.

Now, back to history. About two hundred million years ago, during the age of dinosaurs, Pangaea began to split in two along the Gulf of Mexico and the South Atlantic. The Florida Plateau went with North America. Only, at that time, it was a shallow part of the ocean. As water washed chemicals, minerals, and the shells and bones of sea creatures onto the plateau, limestone formed and eventually emerged from the ocean.

The earth's climate also had an effect on Florida. During the ice ages, when glaciers formed and the sea level dropped, more of the plateau emerged from the sea. When the earth warmed, the ice melted and oceans rose, flooding the plateau. Then more ocean sediment was deposited on the plateau. The land built up in layers, sort of like a sandwich.

During the last ice age, which ended about ten thousand years ago, sea levels dropped so that Florida was twice the size it

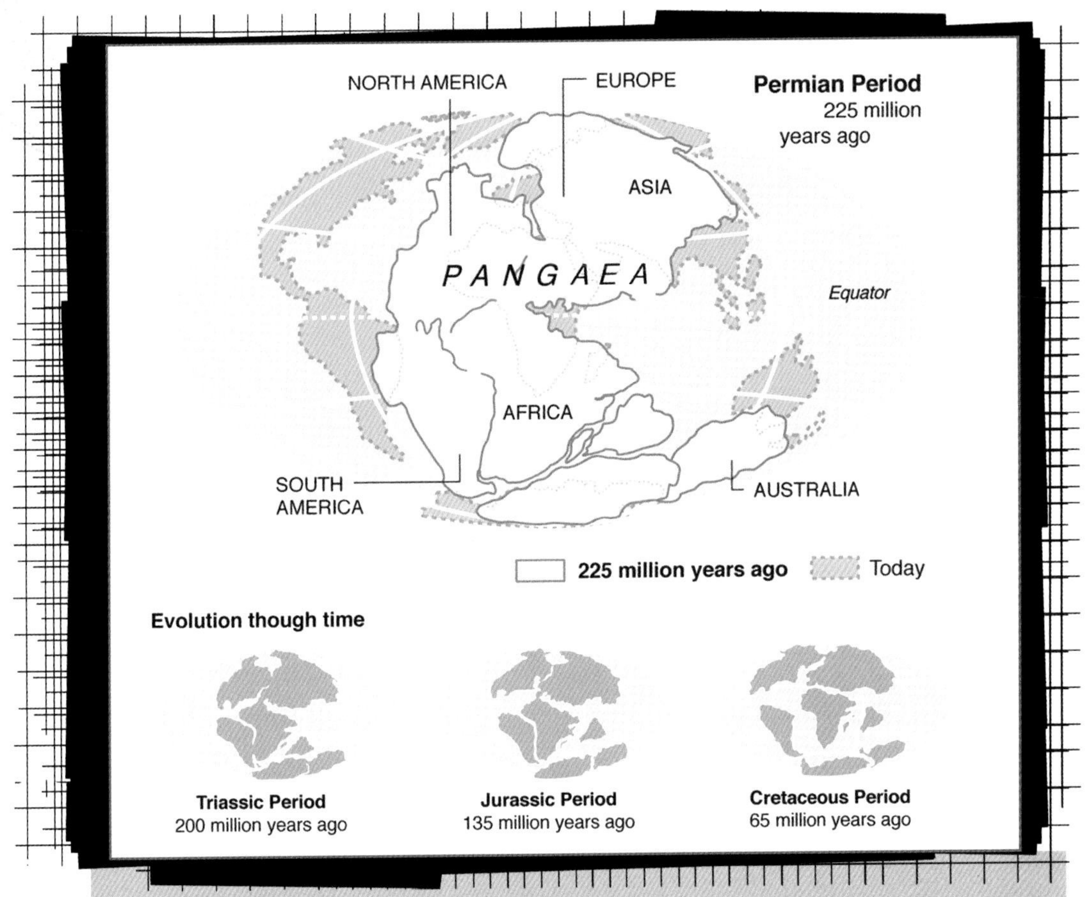

The world was united 225 million years ago as a single continent: Pangaea. When the continents began to split up about 200 million years ago, the Florida Plateau went with North America.

is today. By that time, humans had made their way to North America. Long before they evolved, the continents had broken up into what you see today. So how humans got to the continent is somewhat of a mystery. Many scientists think it was by way of a land bridge connecting what is now Russia and Alaska. An alternate theory suggests that at least some of the earliest Americans got here by boat—either by following the Alaskan shoreline or even by crossing the Atlantic.

Whatever the case, people made their way to Florida and, because of the lower sea levels, hunted on parts of the peninsula that are now underwater. For this reason, scuba divers have found spear points off the coast of Florida.

As the ice age gave way to warmer climates, Florida lost much of its land to the sea. Miami's Biscayne Bay and the Everglades became flooded. Now, the latter is swampland and marshland teeming with life. It's the only place on earth where alligators and crocodiles coexist. Much of the rest of the peninsula became part of the continental shelf, which slopes down 300 feet (91 meters) and then drops off into the deeper sea.

Florida's story shows how many forces of nature can combine to create peninsulas. In this case, continental drift, rising and falling sea levels, and sedimentation all played a part. Other earth processes, such as glaciers and volcanoes, create peninsulas, too. In this book, you'll learn all about that. But first, perhaps you should know more about what a peninsula is . . . and isn't.

The word "peninsula" comes from the Latin *paeninsula*, meaning "almost an island." That makes sense because a peninsula is land that is almost surrounded by water.

Sometimes, an isthmus (a narrow piece of land) connects the peninsula to the mainland. An example of this is the Isthmus of Corinth, which connects the larger Greek peninsula to its smaller southwestern peninsula.

Many isthmuses are world famous because they create shortcuts for seafarers. Historians believe that in ancient times, people would carry their boats across an isthmus, rather than sailing around the larger peninsula. The Isthmus of Kra, which connects Myanmar and Thailand to the Malay Peninsula, was probably one such shortcut. Crossing even a narrow isthmus like this one wouldn't have been easy, however. At its narrowest place, the isthmus is 35 to 40 miles (56 to 64 km) wide, a considerable distance to cross carrying a boat and goods to trade.

Canals created more convenient shortcuts for ships. The Isthmus of Suez connects Africa to the Sinai Peninsula and the rest of Asia. Without a canal, it would also separate the Mediterranean Sea from the Red Sea, which flows into

The Suez Canal cuts through the isthmus between the Sinai Peninsula and Africa. It provides a shortcut for ships, such as the nuclear-powered aircraft carrier shown here, the USS *Nimitz*.

the Indian Ocean. In that case, the shortest shipping route from Europe to the Far East would be around Africa. Today, the Suez Canal creates a shorter route for ships. Contemporary people probably weren't the first to think of this shortcut, however. Archaeologists believe that ancient people constructed a similar canal four thousand years ago.

Not every peninsula has an isthmus. A peninsula can also be a piece of land that juts out into the water without narrowing first. Florida and Italy are examples of this.

A peninsula can even be land that juts out from the mainland and gets narrower in the water, as is the case with India.

Many peninsulas, such as India, create their own country or state. Others share boundaries. The Iberian Peninsula, for example, is home to the countries of Andorra, Portugal, Spain, and the British crown colony of Gibraltar. The Korean Peninsula houses both North Korea and South Korea, countries with very different governments and ways of life.

The above are examples of large peninsulas. But like an island, a peninsula can be almost any size. A smaller peninsula can even jut out from a larger peninsula. For instance, Gibraltar is a smaller peninsula connected by an isthmus to the Iberian Peninsula. It is known for the Rock of Gibraltar, which we will talk about in chapters 4 and 5.

Another type of peninsula is a cape, which is a narrow piece of land jutting into a body of water. Often, it ends at a narrow tip. Sometimes, it curves around and creates a bay. Famous American capes are Cape Cod in Massachusetts, a popular vacation spot, and Cape Canaveral in Florida, where the United States tests spacecrafts. The Cape of Good Hope in South Africa is well

known, too, but it is not the southernmost tip of Africa, as some people think. That would be Cape Agulhas, where the Atlantic and Indian oceans meet.

Now that you know what a peninsula is, here are some things that a peninsula isn't.

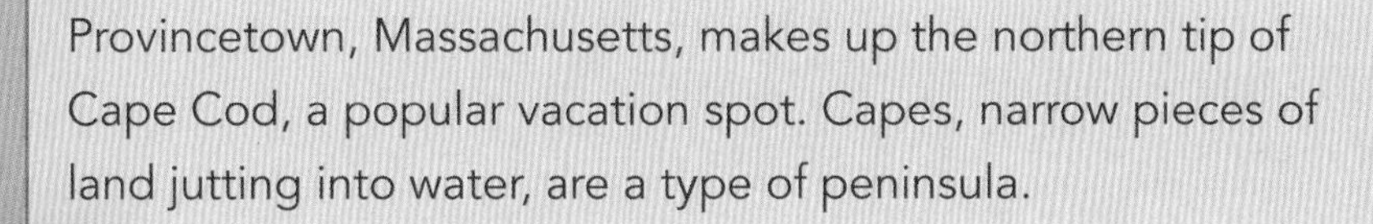
Provincetown, Massachusetts, makes up the northern tip of Cape Cod, a popular vacation spot. Capes, narrow pieces of land jutting into water, are a type of peninsula.

A continent. South America is connected to North America by the Isthmus of Panama (which also has a famous shortcut: the Panama Canal). But neither North America nor South America is a peninsula. They are continents.

Underwater land. If a peninsula is "drowned" by rising sea levels, it is no longer a peninsula. Now it is part of the continental shelf.

Dry land. When sea levels drop or an inland sea evaporates, and the land is no longer surrounded mostly by water, it is also not a peninsula. Now, it is simply higher ground, or a divide. This has happened to the best of peninsulas, including Korea, Italy, Iberia, Greece, and Scandinavia.

WHAT IS THE DIFFERENCE BETWEEN AN OCEAN AND A SEA?

In songs and poetry and books, oceans are often called seas. When you see the two words on a map, however, they mean different things. The ocean is the body of salt water covering about two-thirds of the earth. There are four major oceans: the Atlantic, Pacific, Indian, and Arctic oceans.

A sea is a body of salt water that is partially or fully surrounded by land. The Mediterranean Sea is almost fully surrounded by land. In contrast, the Yellow Sea barely narrows as it flows into the East China Sea. A huge freshwater lake is sometimes called a sea, too.

Geographical names aren't always precise. In truth, people name places what they want, and it sticks. For instance, the freshwater Sea of Galilee is smaller than any of the Great Lakes. And who's to say the Gulf of Mexico shouldn't be called the Sea of Mexico?

The Mediterranean Sea is almost entirely surrounded by land. The North Sea *(above left)*, however, opens into the Arctic Ocean.

An island. When lowlands connect peninsulas to continents, the lowlands are subject to flooding due to rising sea levels. At this point, the highlands become islands. Specifically, they become continental islands, as opposed to oceanic islands. The latter rise from the sea. Continental islands, however, are part of their neighboring continental structure.

A landform that lasts forever. As a little kid, you learned that all living things—whether flowers or bugs or pets or people—die. But you probably thought that nonliving things, such as mountains, islands, and peninsulas, lasted forever. However, by now you know that nothing on earth lasts forever. Mountains crumble. Land washes away. New land is built. Through the years, many peninsulas have formed only to disappear under the sea and reappear later, when sea levels fall. They are constantly changing—usually very slowly.

Water Water Everywhere

Many peninsulas are surrounded by an ocean or sea, but not always. They can also be found in lakes and other bodies of water. Michigan, for instance, is made up of two peninsulas that jut into Lake Michigan, Lake Superior, and Lake Huron.

Peninsulas can be surrounded by more than one body of water, too. The New Jersey Peninsula, for instance, is bordered by the Atlantic Ocean on the east, the Delaware River to the west, and Delaware Bay to the southwest.

Peninsulas often have an impact on the water around them. Because they jut into the water, they can partially enclose it, forming bays, gulfs, or seas. The Korean Peninsula, with China,

Tropical cyclones like Hurricane Rita, shown here, frequently hit Florida. As the warm water from the Atlantic Ocean and Gulf of Mexico rises, cooler air swirls in to take its place.

creates the Yellow Sea. The Arabian Peninsula borders the Red Sea to the west and the Persian Gulf to the east.

Florida, too, helps to create a body of water: the Gulf of Mexico. Spanning 1,000 miles (1,609 km) between Florida and Mexico, it could easily be called a sea. Known for its warm water, it is the starting point of one of the strongest ocean currents in the world: the Gulf Stream.

At the tip of Florida, two warm currents combine with water from the Gulf of Mexico to form the Gulf Stream. The stream shoots through the Strait of Florida and along the East Coast.

As it reaches the Grand Banks in Canada, it mixes with a cold current from the Arctic Ocean and veers east toward Europe. Some people credit the Gulf Stream with making Europe 30 to 40 degrees warmer in the winter, but this is untrue. The current

AMERICAN PENINSULA TRIVIA

1. What three U.S. states make up the Delmarva Peninsula?
2. Which peninsula and state contain the southernmost tip of the continental United States?
3. What U.S. state is made up of two peninsulas—an upper and a lower one?
4. True or false: Baja California is in the United States.
5. On what U.S. peninsula did the Mayflower first land?

Answers: 1. Delaware, Maryland, and Virginia 2. Florida 3. Michigan 4. False—it is part of Mexico 5. Cape Cod

warms both Europe and America during the wintertime, but only by about five degrees. Europeans can thank the Rocky Mountains and winds from the Southwest for the rest.

Peninsulas affect the water around them, and water affects peninsulas, too. That's because it contains sediment, which hardens to become rocks. This is one way peninsulas form.

We've talked about how Florida was formed by sedimentation in the ocean. In this chapter, you'll learn more about sedimentation and how it occurs both under the sea and on land. We'll look at two different Gulf Coast land areas, Florida and coastal Louisiana, which are examples of each.

Basically, sedimentary land forms when sediment (whether from rock fragments or the remains of living things) is deposited and buried by more sediment. Compressed by the weight above and cemented together by minerals, it becomes sedimentary rock.

On land, sedimentation begins with an opposite process: erosion. In Latin, "erode" means "to gnaw," and that is a pretty good description. Erosion occurs when water, wind, or ice loosen soil and rock from the earth's surface and dissolve it. Then, the sediment is carried someplace else.

Think about a river. As it rolls along, it erodes the soil and rock in its path, depositing it along the way and at its endpoint. Often, at the mouth of the river, the sediment builds up to form a delta, a landform shaped like a fan. You see this along the coast of Louisiana in an area called the Mississippi River Delta and Wetlands. Here, the Mississippi River meets the

At Horseshoe Bend in Arizona, the Colorado River eroded the sandstone around the higher ground and formed a peninsula.

Gulf of Mexico. Over thousands of years, the river formed several deltas, making the coast what it is today. Rivers choose the path of least resistance to get to the sea. Through the years, the Mississippi River took the easiest route to the gulf. As it deposited sediment into the shallow coast along that route, the land built up, becoming either dry land or marshland. Both slowed the flow of the river. Never one to work harder than it needed to, the river found an easier path to the sea. Along the new path, it formed a new delta. That happened multiple times.

In its wake, the river left abandoned deltas, which are now wetlands, all along the coast of Louisiana. The region resembles a peninsula—or a series of peninsulas—in that several areas of land are surrounded almost entirely by water. But it is called a delta.

Today, the Plaquemines-Balize Delta is the main place where Mississippi River sediment is deposited. Because the sediment here is being deposited into deep water, it would need a large

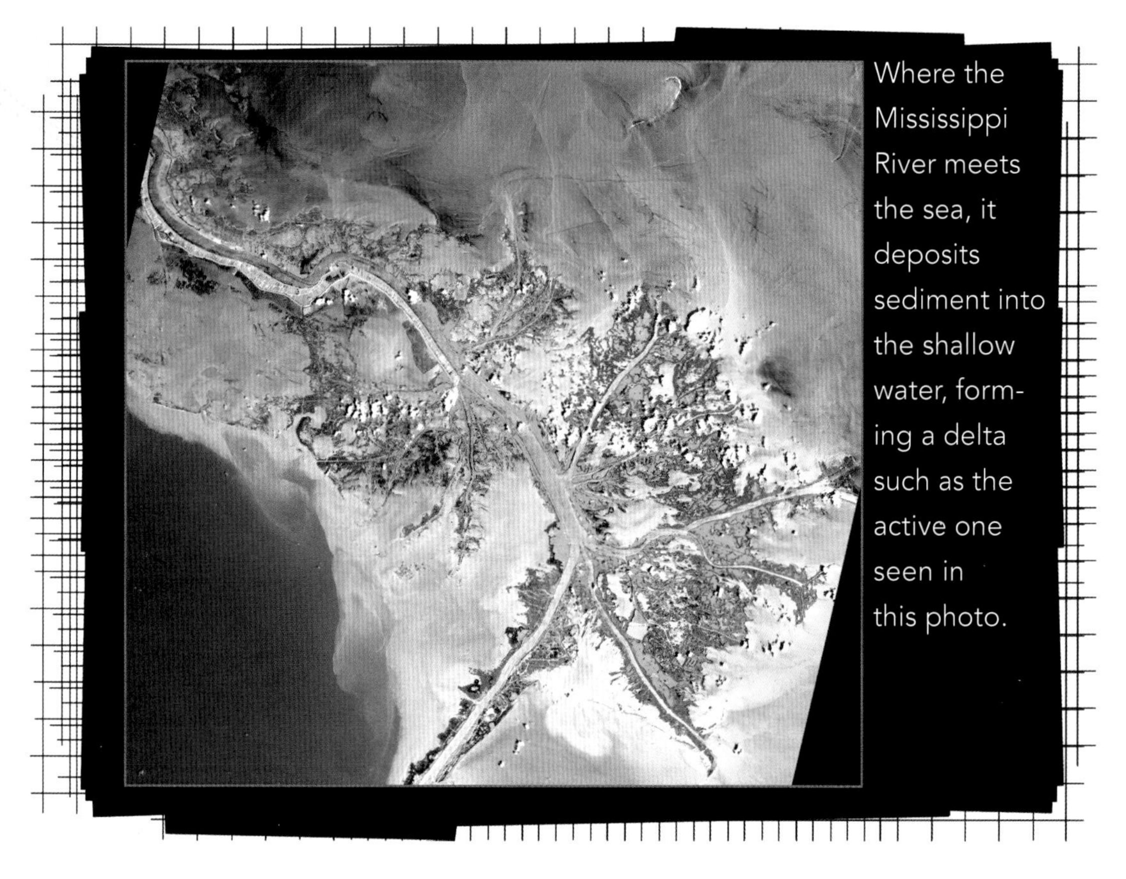

Where the Mississippi River meets the sea, it deposits sediment into the shallow water, forming a delta such as the active one seen in this photo.

amount to add to the land faster than the Gulf of Mexico is eroding it. Even though the Mississippi River carries, on average, 436,000 tons of suspended sediment loads per day, it's not enough. Land is being lost.

Living Rocks

A different kind of sedimentation occurs in the ocean. When shellfish, coral, microscopic plankton, and other sea creatures die, they sink to the ocean floor. Minerals help to cement these things, and they harden to form limestone. If you look closely at

WHAT'S THE DIFFERENCE . . .

. . . between a gulf, a bay, and a lagoon?

A bay is a part of an ocean, sea, or lake that cuts into the shoreline. It is partially enclosed by land. A gulf is the same thing, only much larger.

A lagoon is a shallow body of water that is related to a larger body. It can be an alcove in a lake, for instance, or a saltwater pond separated from the ocean by sand dunes.

. . . between a swamp, a marsh, and a bog?

All are shallow water areas. A swamp contains trees or shrubs. A marsh has neither but is filled with plants like grass and cattails instead. A bog is typically found in colder climates and is made of peat, which is slowly decomposing plant matter.

limestone, you'll see fossils of shells and other living things. (You can do this even if you don't live near an ocean; limestone found in the Midwest is rich with fossils from when the land was part of an inland sea.)

While limestone forms under the sea, it is very plentiful on dry land. Limestone emerges from the sea in many different ways. If it forms in a shallow part of the ocean, it can ultimately rise above sea level. Alternately, the ocean floor can be pushed up to sea level by processes happening inside the earth. Falling sea levels can also expose limestone. Or, as in the case of the Midwest, the sea may cease to exist entirely, exposing the limestone structures it created.

Not all limestone is alike. It differs in appearance, composition, and age. Several different types of marine limestone make up Florida. Ocala limestone underlies the peninsula and dates back to about thirty-eight to fifty-four million years ago. It is made up of shells and chalky material. In contrast, Miami limestone, found at the surface of southeast Florida, is much younger. It was formed during the Pleistocene epoch 10,000 to 1.8 million years ago. It is partly made up of round grains that look like fish eggs. Both types of limestone contain fossils.

Limestone makes up many other peninsulas, including Italy. It, too, was formed, in part, by sedimentation from the Tethys Sea. Never heard of it? That's because it no longer exists—not in its historic form, anyway. But for millions of years, it was one of our planet's major seas.

Even before Pangaea, there was a Paleo-Tethys Ocean. As the continents moved, that ocean became the site of the Tethys Sea. It, too, changed through the years but was roughly located north and east of Africa, south of Eurasia, and north of India (while it was still a separate landmass south of Asia). When India traveled north and collided with Asia fifty million years ago, some say it marked the closing of the Tethys Sea. However, the Tethys Seaway still existed in the region of the present-day Mediterranean, linking the Atlantic and Indian oceans. The seaway closed much later, fourteen to eighteen million years ago, when Arabia and Iran collided with Asia, dividing the Indian Ocean from the former Tethys. Now, the Mediterranean is a remnant of the Tethys Sea.

During its heyday, the Tethys Sea covered much of present-day Europe, including Italy. Sediment from the Tethys built up the peninsula, just as Florida was built up by marine sedimentation. In Italy, the sediment formed shale, sandstone, claystone, mudstone, and limestone.

What is now the Wadi Hatin World Heritage Site in Egypt was once at the bottom of the prehistoric Tethys Sea. Now on dry land, the seabed has been eroded by wind, creating these interesting landforms.

Then, another force of nature helped to form the peninsula. Tectonic plates collided, putting pressure on underwater sedimentary rock and causing it to pile up like a row of blocks pushed together from either side. This formed the Apennines, the mountain range that, today, runs the length of Italy.

Next, sediment from the land and sea played a role in the formation of the peninsula. As sea levels rose, the Tethys flooded the slopes of the mountains, and marine sediment built up. When the sea level dropped, land sediment from the crumbling mountain formed new rock. Because of this, Italy is comprised of both marine and land (or continental) sedimentary rock.

OCEANFRONT PROPERTY IN KANSAS

When you think of the seashore, you probably picture Florida, California, or maybe a beach close to your home. If you'd lived about ninety-four million years ago, however, you might have taken a seaside vacation to Kansas, Nebraska, or someplace else in the Midwest. At that time, the Western Interior Sea, which was about 600 feet (183 m) deep, covered the area.

This sea wasn't for swimming, though. It was home to sea monsters like the mosasaur, which grew as long as 50 feet (15 m) and had an insatiable appetite for smaller creatures.

By about sixty-five million years ago, many animals in this sea had become extinct. Soon, the sea disappeared, too. When the Rocky Mountains rose, they lifted the seabed with them. The Western Interior Sea closed, ruining vacation plans for future Midwesterners everywhere. However, you can still see fossils of sea creatures in the Midwest limestone.

The Italian Peninsula was formed by a variety of processes, including sedimentation from the sea and the creation and erosion of mountains. (Erupting volcanoes even played a role.) As you read on about peninsulas, keep in mind that while one prevailing reason might have led to the formation of a peninsula, a variety of the earth's processes probably played a part.

It's also important to remember that the earth is in constant flux. The peninsulas of today could be part of the continental shelf tomorrow and vice versa, or they could be inland one day.

Several geological processes, including marine sedimentation, contributed to the formation of the Italian Peninsula, home of the seaside town of Vernazza, shown here.

What scientists know about the earth is also changing, and many geological theories are relatively new. For instance, the theory of continental drift was proposed—and widely rejected—less than one hundred years ago. Today, scientists accept that the continents move, but they are still studying how and why they've drifted through the years and continue to drift. One thing is certain: nothing on earth is constant—except change, that is.

You hear a lot today about global warming, but it's not a new phenomenon. Throughout history, the earth has warmed and cooled. The differences now are that, a) scientists believe humans are causing the rapid warming trend, and b) the earth is warming at the fastest rate in millions of years.

About fourteen thousand years ago, the earth was also warming, but for natural reasons. It was emerging from the latest ice age. A glacier that had crept across much of North America, Europe, and Asia was melting. The ice age and subsequent meltdown created peninsulas in several different ways.

Before we talk about that, let's look at how glaciers shape the land. First, picture midnight snow falling on your town. You probably don't expect that snow to go anywhere. It will stay right there until it melts a few days later. Well, glaciers are not like that at all. Rather, as the ice gets up to a mile or two deep, the sheer weight of the ice on top pushes the ice below, causing it to flow downhill. (To see this phenomenon for yourself, try the experiment featured in this chapter.)

Second, glaciers are not made of pure ice, like the cubes you put in your cola. Instead, they contain sand and pebbles and

This ice mushroom formation, part of the Nuptse Glacier on Mount Everest, illustrates how glaciers carry not only fine sediment but also rocks and boulders.

even boulders. As the glaciers move, this sediment scrapes the land, eroding it, particularly if it is made of soft rock.

Finally, glaciers melt, depositing the eroded rock and soil onto new places, forming lakes, and releasing water back into the oceans, causing the sea level to rise.

Several ice ages occurred during the Pleistocene epoch, which was 1.8 million to 10,000 years ago. During this time, the temperature sometimes warmed. Then it cooled again. These periods of time, lasting thousands of years, were called glacial (or ice age) and interglacial periods. The last glacial stage was the

GO WITH THE (GLACIAL) FLOW

To see how glaciers flow, try this experiment.
You'll need: A square cake pan, water, a larger tray or pan, a brick, and a ruler.

Directions:

- Completely freeze water in the cake pan.
- Carefully remove the ice by running warm water over the pan. Place it on the larger tray or pan.
- Measure the length of the ice.
- Place a brick on top of the ice.
- Put the ice back in the freezer.
- After one or two days, measure the ice again.

Is it longer than its original length? It should be. It might have even expanded to fill the larger tray. The weight of the brick pressed on the ice, causing it to flow without melting. During the ice age, the weight causing the glaciers to flow was actually the glaciers themselves. The ice on top weighed down the ice at the bottom, causing it to move like the ice underneath your brick.

Wisconsin, approximately seventy-five thousand to ten thousand years ago. At the end of the Wisconsin, the world warmed, marking the end of the Pleistocene epoch.

Glaciers created peninsulas during and after the Wisconsin stage in several ways.

Erosion In Wisconsin and Michigan, three landforms, now called the Keweenaw, Door, and Bayfield peninsulas, were formed by glacial erosion. Remember how glaciers scrape across the land, eroding soft rock? Well, these three areas were not made of soft rock, but of highly resistant rock. As the glacier flowed downhill, it eroded the soft rock, which would eventually become the basins of Lake Superior, Lake Michigan, and Green Bay, but left the stronger rock intact. When the glaciers melted and the lakes formed, the rock that was not eroded was on higher ground and therefore became peninsulas.

Meltwater Like water, glaciers flow downhill, moving fastest along the paths of least resistance, such as river valleys. Often, glaciers widen and deepen these areas—or erode them, as we talked about above. This process, in part, is what carved the basins for the Great Lakes.

Next, when the glaciers melt, they deposit the dirt, rocks, and sand they carried with them. Large deposits are called moraines, which are hills or ridges. In some cases, the moraines act like natural dams for the glacial meltwater. An earlier version of Lake Michigan, for example, pooled behind a moraine to its south. Lake Huron formed in a similar way. These two lakes surround what is now referred to as Michigan's Lower Peninsula.

Deposition On the East Coast, the North American glacier of the Wisconsin stage extended to Massachusetts and what are now the islands of Martha's Vineyard and Nantucket. These are just south of the Cape Cod Peninsula. Here, the edge of the glacier reached its maximum about twenty-one thousand years

ago. Then, as the earth warmed, the glacier melted faster than it flowed. It retreated north but stalled for a while at what is now Cape Cod. It flowed and melted at the same rate, sort of like it was on a treadmill. Because of this, lots of glacial drift was deposited in one place, creating Cape Cod.

Most glaciers, such as the Kennicott Glacier in Alaska, exist toward the North and South poles. During the last ice age, their icy fingers stretched south over several present-day American states. The sediment-laden ice created many peninsulas.

At this time, the land area of Cape Cod was much greater because the sea level was about 300 feet (91 m) lower. Think about rowing a boat to where the ocean is 300 feet deep—you'd be pretty far offshore. Thousands of years ago, your boat would have still been on dry land. Nantucket, Martha's Vineyard, and Cape Cod Bay were all dry areas, except for when a glacial lobe melted, forming a lake in what is now Cape Cod Bay. The lake eventually drained into the sea.

Then, by six thousand years ago, the sea level had risen almost 300 feet. Cape Cod Bay flooded again, this time with ocean water. So did Nantucket and Vineyard sounds. The peninsula began to look how it does today. However, if the sea level continues to rise at its present rate of 3 feet (1 m) per year, Cape Cod will be under the sea within five thousand years.

Rising Sea Levels As you can tell from the story of Cape Cod, rising sea levels play a big part in the formation of peninsulas. Next, you'll read about some of the peninsulas caused by floods, the result of rising sea levels.

From Sea to Rising Sea

You learned that Florida was a larger peninsula when ice age sea levels were lower than they are today. In North America and throughout the world, many peninsulas were not peninsulas at all during the ice age.

Today, for instance, the Korean Peninsula is nearly surrounded by the Yellow Sea to the west, the East China Sea to the south, and the Sea of Japan to the east. But during the ice age, the Yellow Sea was simply a low-lying area of land next to the Pacific Ocean. When the ice melted and sea levels rose, the lowlands were flooded and became the Yellow Sea.

Other highlands that became peninsulas when lowlands were flooded include Scandinavia, Italy, Alaska, and Arabia.

During the ice age, the Arabian Peninsula still met the Red Sea to the west and the Indian Ocean and Arabian Sea to the south. But to the east, the Persian Gulf was land that connected Arabia to present-day Iran.

Alaska was connected to Asia by a land bridge. Many scientists believe an ice-free corridor on the land bridge was the route humans migrating from Eurasia to North America took. By about

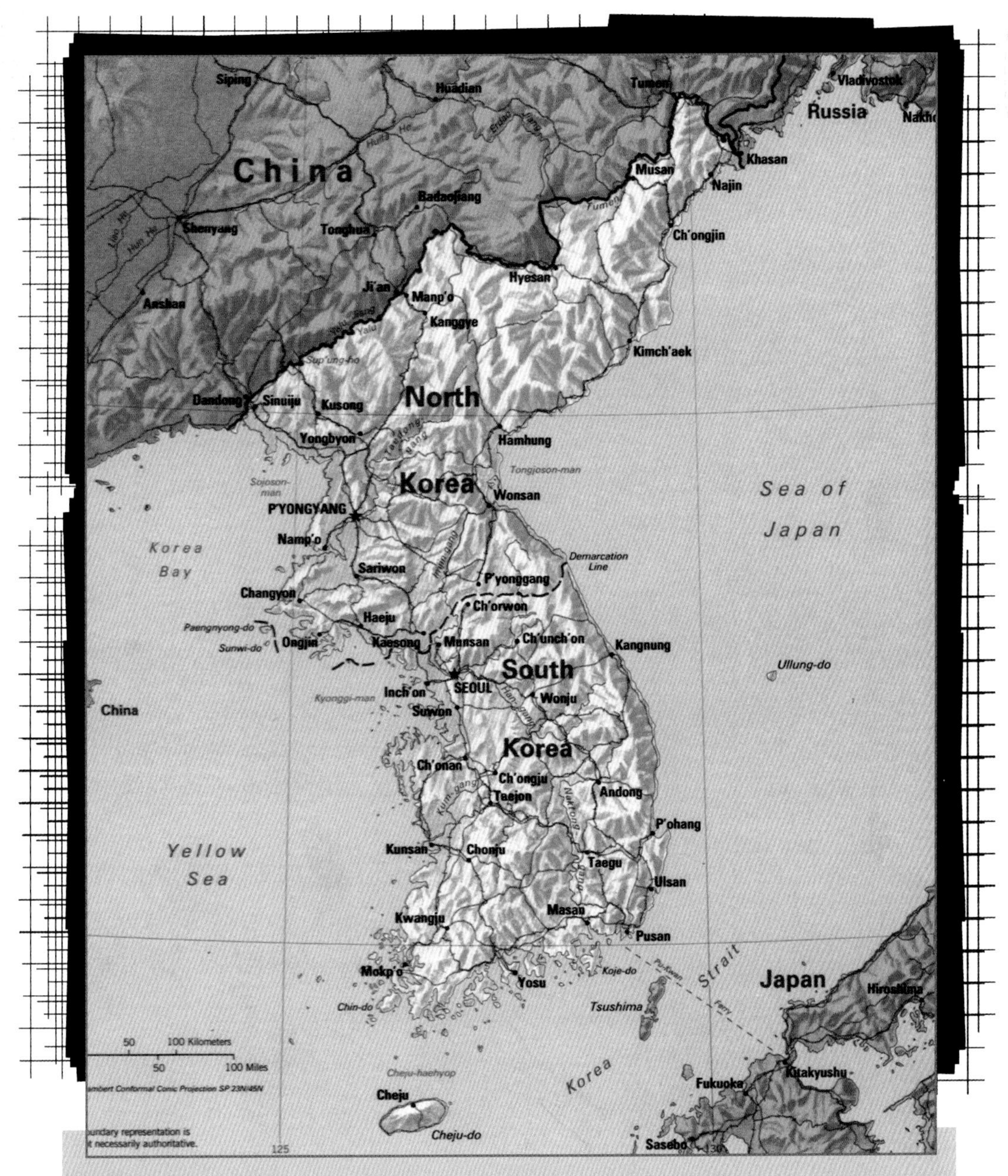

During the last ice age, many peninsulas didn't exist. With water trapped in glaciers, sea levels dropped. The Yellow Sea was dry land.

ten thousand years ago, this land bridge was underwater, cutting off any route to North and South America other than by sea.

While many highlands became peninsulas because of rising seas, other peninsulas at the time became islands. As the sea level rose, the lowlands between the peninsula and the mainland flooded. Most of the Indonesian islands, for instance, were parts of peninsulas attached to mainland Asia or Australia before the deluge.

By the time the ice age ended, people had spread to every non-glaciated corner of the earth. They endured frequent flooding due to the melting ice. Scientists, historians, and archaeologists believe these floods led to the creation of several legends. From Noah's Ark to an Irish legend of a queen sailing for seven years as oceans flooded Ireland, scholars think ancient people's flood stories referred to either a post–ice age deluge or, more likely, the frequent flooding of their hunting territories.

Today, scientists predict that if global warming continues, and Greenland's ice sheet melts, sea levels will rise 23 feet (7 m). Just as the end of the ice age shrank Florida's landmass, so would the melting of Greenland submerge some of the Florida Peninsula, including Everglades National Park and Miami. While it would not cause a rise in sea level of post–ice age proportions, global warming could cause terrible floods for coastal cities and nations throughout the world.

CHAPTER 4

COLLIDING, ERUPTING, GUSHING: DRAMATIC PENINSULA FORMATIONS

You've learned about two common ways peninsulas form: through sedimentation and glaciation (including the rising sea levels that followed the ice age). But the truth is, peninsulas form in almost as many ways as every other landform on the earth forms. Mostly, they form slowly and subtly. Occasionally, dramatic events, including clashing continents, gushing waterfalls, and exploding volcanoes, have led to the creation of peninsulas.

The Rock of Gibraltar

If you've ever heard somebody say "like the Rock of Gibraltar" in a sentence, he or she is probably talking about a person or thing that is strong and reliable. It's a well-known symbol. An insurance company uses the landform in its logo, and the rock has been featured in love songs by artists ranging from 1950s icon Frankie Laine to today's alternative rock singer Nick Cave.

The rock became famous for many reasons. First, measuring 3 miles (4.8 km) long, three-quarters of a mile (1.2 km) wide, and 1,396 feet (425.5 m) at its highest point, Gibraltar is a big rock! Second, it's riddled with caves, once providing natural shelter and hideouts for people

Gibraltar is a peninsula attached to another, the Iberian Peninsula. Many view the enormous Rock of Gibraltar, seen here, as a symbol of strength. The formation also has an interesting geological history.

and Neanderthals. Third, it marks the only entrance from the Atlantic Ocean into the Mediterranean Sea, making it a strategic military post through the years. The waterway connecting the Atlantic to the Mediterranean is called the Strait of Gibraltar.

Geologically speaking, what makes the rock so interesting is how it differs from the rest of the landscape on the Gibraltar Peninsula. The rock is made of gray limestone, probably dating back to the Jurassic age. In contrast, the land surrounding it is flat, much younger, and covered with sand 30 feet (9 m) deep.

There's a reason why the rock is so different from the rest of the peninsula: two hundred million years ago, its limestone mass

PILLARS OF HERCULES

Together, the Rock of Gibraltar and Mount Ceuta in Morocco are called the Pillars of Hercules. You've probably heard of the mythical Greek hero Hercules. Known for his strength, he strangled two giant snakes while he was still a baby. As a boy, he killed a lion with his bare hands. Then, when he got older, Hercules went crazy and did something terrible.

To atone for this, he was given twelve tasks—often involving slaying a beast with multiple heads. His tenth task, for example, was to capture the oxen of a three-headed monster. On his way back from doing that, he celebrated his journey by placing two rocks—mountains, actually—on either side of the Strait of Gibraltar.

Whether or not you believe that story to be 100 percent true, it shows that the ancient Greeks traveled through the Strait of Gibraltar and considered the two rocks to be historically significant, as we still do today, but in a different way.

grew under the ocean. Like Florida and Italy, it was created by the remains of sea animals. Then, Africa crashed into Europe, dislodging the rock and throwing it westward, where it came to rest on the southern edge of the Iberian Peninsula. It didn't rest for long, though. The colliding continents put pressure on the rock until it flipped over. Today, Gibraltar is an upside-down rock.

The rock hasn't always been a peninsula. Across the Strait of Gibraltar, there is a sill—a natural dam. As long as the Atlantic Ocean rises above this sill, it continues to replenish the Mediterranean. But if sea levels were to drop below the sill, the Mediterranean would become cut off from the Atlantic and evaporate. In that case, neither Gibraltar nor Iberia would remain a peninsula. Instead, they would be bordered on the east by a gaping hole six times the size of California.

About six million years ago, this exact thing happened. Falling sea levels and pressure from the ongoing collision of Africa and Europe caused the sill to dam the Atlantic. The Mediterranean dried up. Over the next million years or so, it sometimes refilled partially, only to evaporate again.

Then, because of rising sea levels and possibly other natural causes, the Atlantic breached the natural dam. An amazing waterfall—gushing at ten times the rate of Niagara Falls—spilled over the sill and dropped to the depths of the dry Mediterranean Sea, filling it in a very short geological time. Gibraltar became an island, and as sea levels dropped, the island became a peninsula.

Today, Gibraltar is surrounded by the Mediterranean Sea and the Strait of Gibraltar. It is connected to another peninsula, the Iberian Peninsula, which is bordered by the Mediterranean, the Strait of Gibraltar, and the Atlantic Ocean.

Scientists believe that as Africa continues to push north, the Mediterranean Sea will again be replaced by dry land—this time by a mountain range pushed up by the collision.

Crashing and Burning

The Gibraltar Peninsula became what it is today, in part, because of colliding continents. Likewise, India—which was a separate landmass long ago—became a peninsula when it crashed into Eurasia. But first, India made a long journey.

It was a lone landmass off the coast of Australia 225 million years ago. Then, when Pangaea split apart about two hundred

WORLD PENINSULA TRIVIA

1. What is the Italian Peninsula shaped like?
2. What volcano erupted in 79 CE, burying the Italian town of Pompeii?
3. The Sinai Peninsula is part of what country?
4. Which cape marks the southernmost tip of Africa?
5. What countries make up the Malay Peninsula?
6. What Mexican peninsula is surrounded by the Gulf of Mexico and the Caribbean Sea?

Answers: 1. A boot 2. Vesuvius 3. Egypt 4. Cape Agulha 5. Myanmar, Thailand, and Malaysi 6. Yucatan Peninsula

million years ago, India began its journey north of thousands of miles. Scientists have learned that eighty million years ago, India was 3,977 miles (6,400 km) south of Asia and traveling at about 9 meters (30 feet) per century.

By footrace standards, India wasn't breaking any speed records. If you traveled at a rate of 9 meters per century, for instance, it would take you ten years to take just a couple of steps. By geological standards, though, India was moving at a record-breaking speed. Its arrival had dramatic effects, too. As it collided with Eurasia forty thousand to fifty thousand years ago, it continued to advance, leading to the formation of the Himalayas, a mountain range 1,802 miles (2,900 km) across and with peaks 5.6 miles (9 km) high.

When India collided with Eurasia, India became a huge shield-shaped peninsula. Pressure from the collision, illustrated by the arrows in this diagram, formed the Himalayas.

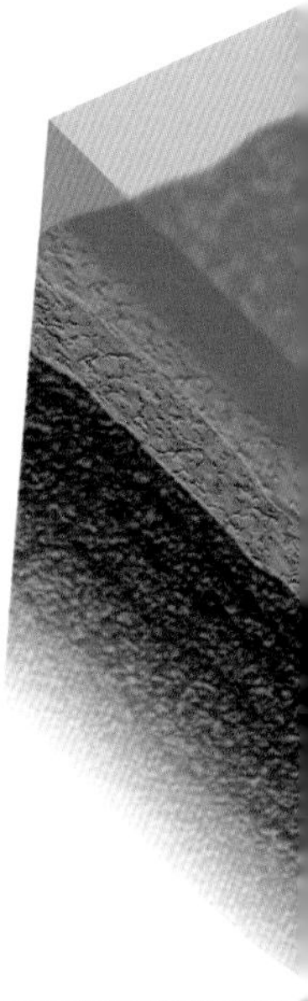

India is no longer alone but is a peninsula in Asia. No human being was on the earth to witness the collision, but even today, people experience its effects. As India's tectonic plate continues to push north, it puts pressure on Asia, leading to terrible earthquakes along the continent's fault lines.

The collision of continents can also cause peninsulas to become something else. Panama used to be a peninsula at the southern edge of Central America (which was part of North America). When North and South America collided, the peninsula became an isthmus connecting the two American continents.

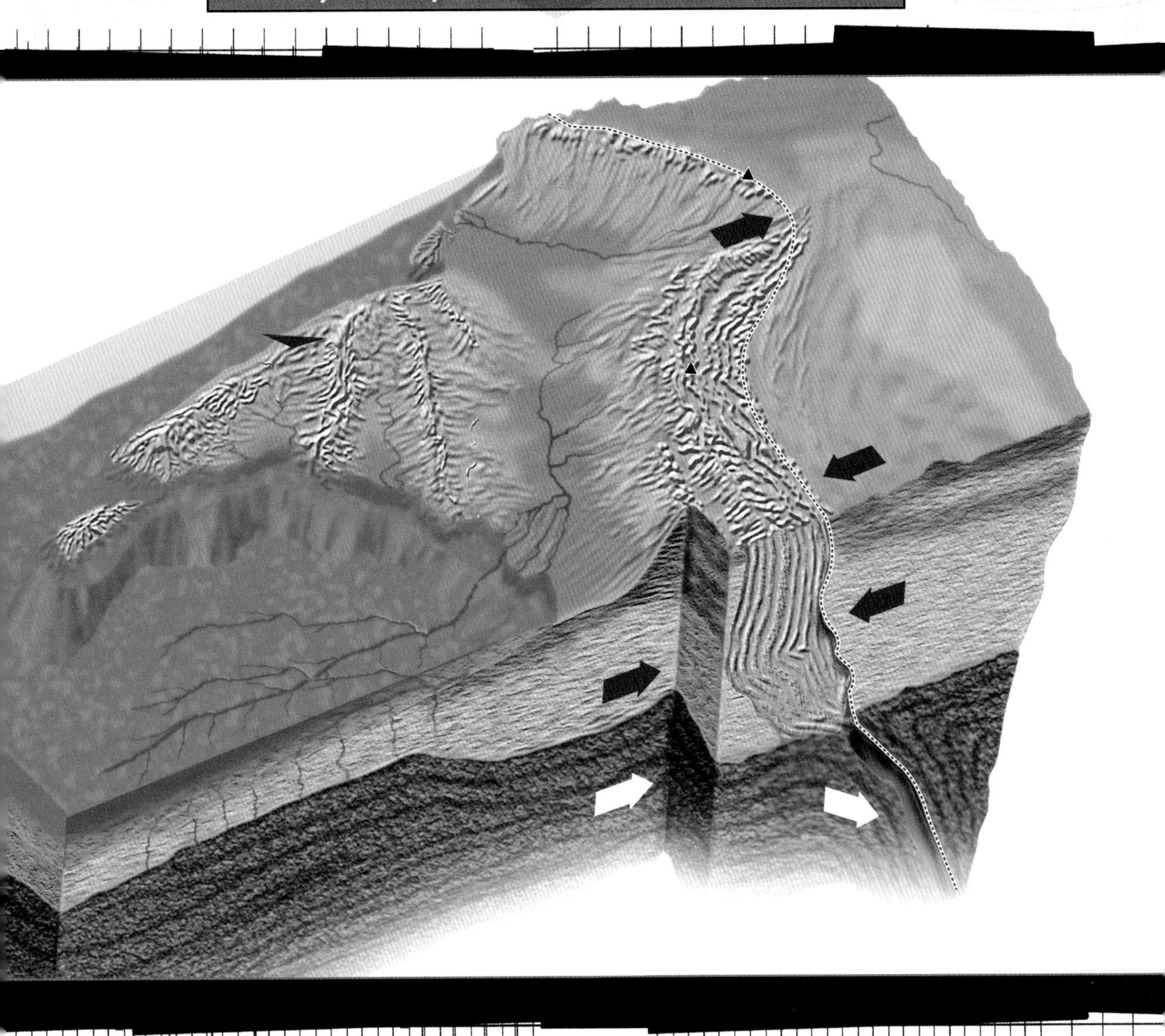

Another way plate tectonics form peninsulas is by separating land. About forty million years ago, Arabia was completely connected to Africa and was part of the African Plate. Then, by about thirty-eight million years ago, a rift occurred. A new plate formed—the Arabian Plate—which began moving north. Now, it is a peninsula bordered by the Red Sea, the Gulf of Aden, the Arabian Sea, the Gulf of Oman, and the Persian Gulf. It is made up of the countries of Jordan, Iraq, Kuwait, Bahrain, Qatar, the United Arab Emirates, Oman, Yemen, and Saudi Arabia.

Finally, volcanic eruptions can create peninsulas, as with the Peninsula de Aseses in Nicaragua. In prehistoric times (but within the last ten thousand years), the Mombacho Volcano erupted violently, causing a terrible avalanche of mud and rock. The avalanche traveled 7.45 miles (12 km) and landed in the shallow shoreline waters of Lake Nicaragua, forming islands called Las Isletas de Granada and the arc-shaped Peninsula de Aseses. Peninsulas like this are common when volcanoes erupt near shallow bodies of water.

People now live on the peninsula and islands. While the volcano hasn't erupted for five hundred years, scientists fear that an earthquake or hurricane could start another debris avalanche that would endanger the people living there.

CHAPTER 5

MAKING WAVES: PEOPLE AND PENINSULAS

Since Neanderthal times, people have sought shelter on peninsulas. It makes sense. Living near the sea—whether on a peninsula or other coastal area—allows people to diversify their food options. Neanderthals living on the Gibraltar Peninsula, for instance, dined not only on land animals like rabbits and tortoises, but also on mussels, dolphins, and seals.

The Rock of Gibraltar, in fact, was the perfect place for cavemen. In addition to the diverse foods available, it had so far escaped the icy climate that gripped northern Europe. For shelter, it had caves, where families gathered around fires. Today, those caves meet the shoreline. At the time, however, with sea levels lower due to glaciation, a couple miles of savanna protected the caves from the Mediterranean Sea. That's important because living right next to the coast would have made people susceptible to wind and storms.

Gibraltar was perfect for Neanderthals, that is, until the ice age stretched its frosty fingers to the south of Spain. Then, the climate, competition with modern humans, or some natural event—scientists don't know exactly what—took its toll on the small number of cavemen living there. They died out. Gorham's Cave, where a hearth has been dated

to twenty-eight thousand years ago, is the last known living place of Neanderthals.

For a long time, modern humans had shared parts of Eurasia, including the northern Iberian Peninsula, with the Neanderthals. But by about twenty thousand years ago, modern humans had completely replaced Neanderthals on the Rock of Gibraltar.

Over the next thousands of years, modern humans migrated around the world. They began to trade with one another. As people developed trade routes, peninsulas played important roles because of their proximity to the sea. Along these routes, both goods and ideas were exchanged. The Greeks, for instance, imported the knowledge of Egyptian art, Mesopotamian mathematics

Gorham's Cave, located on the Gibraltar Peninsula, is the last known home of Neanderthals. Because of diverse food sources and convenient trade routes, peninsulas have always attracted people to their shores.

and astronomy, and the Phoenician alphabet to develop a highly intellectual society. In turn, they philosophized, wrote literature about god and man, instated the first democracy, and held the first Olympic games. Those ideas spread, enriching the rest of the world.

Other peninsulas became places of cultural and religious importance, too. The Sinai Peninsula is holy land for Jews, Muslims, and Christians. India gave birth to both Buddhism and Hinduism. The Italian Peninsula is seat of the Roman Catholic Church. Italy also gave rise to the Roman Empire and, later, the Renaissance era.

While they have been hubs for commerce and cultural exchanges between nations, peninsulas also have natural boundaries that can

DO YOUR PART

Peninsulas represent some of the most beautiful seaside and lakeside places on the earth. Whether you live by the ocean or in the middle of America, you can help protect them. Here's how:

- Pick up trash on the beach, beside rivers, or on your school playground. (If it gets in a storm drain, it gets in the rivers, and eventually into the ocean.)
- Tap water comes from rivers, lakes, and oceans. Don't waste it!
- Learn about global warming and help curb climate change.

Here are just a few ideas:

- Ride your bike, rather than riding in a car.
- Turn off your computer at night.
- Turn off lights when you leave the room.

separate them from other political entities. Often, a country marks its boundary along the isthmus of a peninsula or where the land juts into the sea, as is the case with Italy. Other times, peninsulas are either part of a larger political territory or are divided into more than one country. This can create tension, especially if the countries differ ideologically, as is the case with North Korea and South Korea.

Historically, the Korean Peninsula was made up of both Chinese and Korean kingdoms and had ties to Japanese kingdoms, too. By the time World War II was fought, however, Japan had annexed the country. During the war, both Russia and the United States fought Japan for its land. When the country surrendered, Korea was divided between the two new world powers. It was occupied by America to the south and Russia to the north. Eventually, the peninsula was to be united into one country.

That didn't happen. Instead, North Korea became a Communist state that has struggled economically. South Korea developed a free market and has fared better. Tensions between the two countries have eased, but people living in the north still suffer because of the poor economy and totalitarian government.

Population and Pollution

In early times, people didn't have a huge impact on the land. They may have built canals, farmed once-wild fields, and even possibly hunted some animals to extinction. More than anything, though, they were changed by the earth—not the other way around. (Some scientists, for instance, think that the harsh conditions of the ice age forced humans to evolve into the creative-thinking people that we are today.)

Since contemporary times, however, people have affected the land very much—with both good and bad results. Perhaps the greatest impact people have had on a peninsula can be seen in Boston. It used to be a peninsula. Now it isn't.

When immigrants first settled Boston in 1630, it was a 783-acre peninsula connected to the mainland by a narrow isthmus. On the peninsula were three small mountains: Trimount, Copps Hill, and Fort Hill. The city stayed that way for more than one hundred years, and the small peninsula was sufficient for the

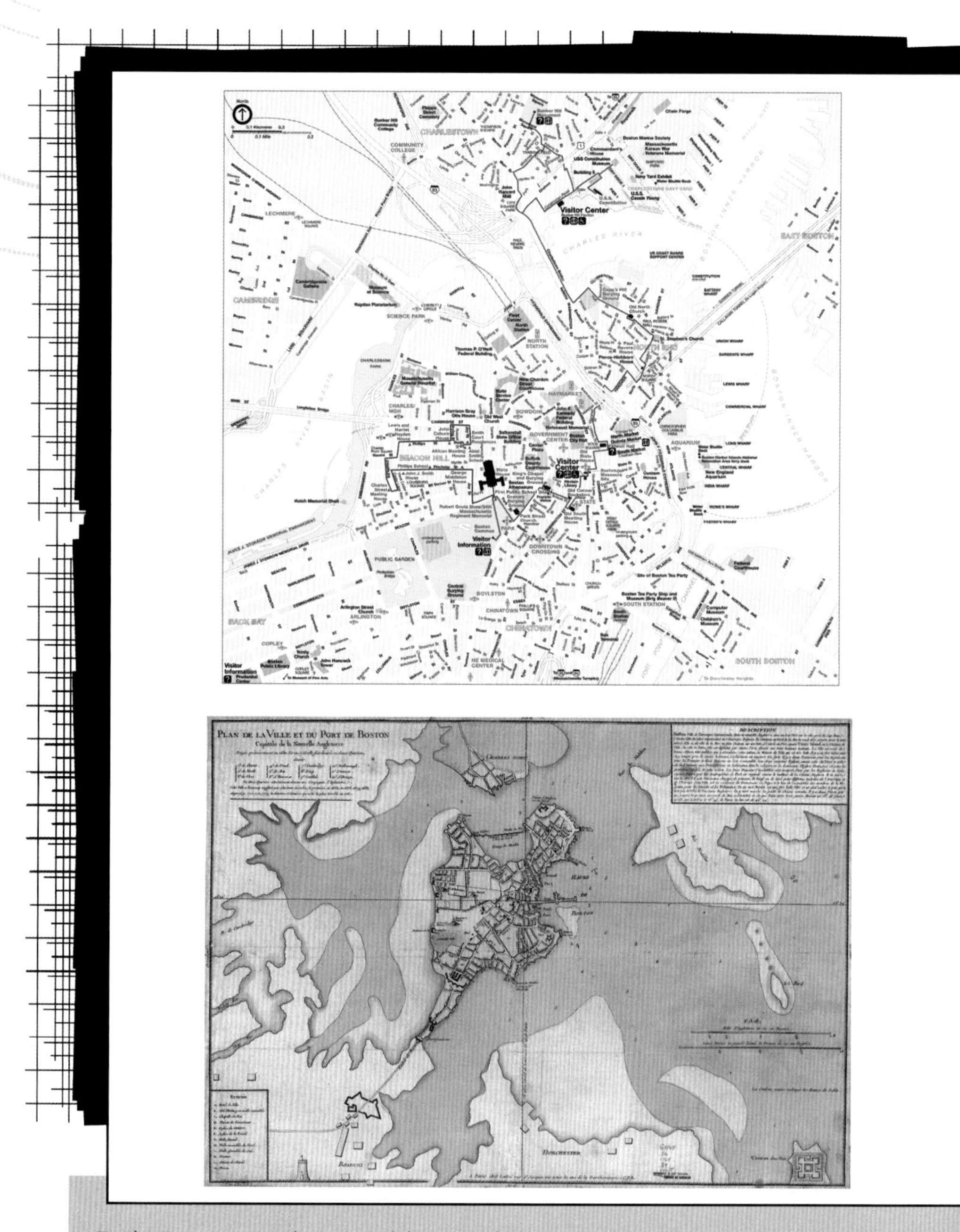

Early Boston settlers would scarcely recognize the city today (*top*). Landfill projects dating to the 1800s transformed what was the Shawmut Peninsula (*bottom*) into a much larger area.

sixteen thousand to eighteen thousand people living there at the time.

However, in the 1800s, the population of Boston exploded. Seeking a better life or fleeing a horrible one, immigrants from Europe traveled across the Atlantic to America. Many settled in cities on the East Coast, such as Boston.

Soon, Boston had outgrown what had been known as the Shawmut Peninsula. Boston developers came up with an ambitious plan. They would move their mountaintops. Using horses and carts, they moved land from Trimount to fill three coves. Other coves, South Boston, and the South Bay would follow.

People continued to flood Boston. (By 1875, Boston's population would be 347,000, or nearly twenty times what it had been less than one hundred years earlier.) At the same time, an attempt to dam the city's Back Bay in order to harness power had failed. Instead, citizens were left with a stagnant, polluted pond. Something had to be done.

Developers began the biggest landfill project of all. The Back Bay was 700 acres, about the size of the original Boston Peninsula. The city was already built, so "spare mountains" were no longer available. But the emergence of the railroad and steam shovel allowed workers to bring landfill from farther away. Every day, they transported 3,500 cartloads of gravel to the Back Bay. It took thirty-seven years to fill the bay. When the project was complete, it became home to the Boston Public Library, the Museum of Fine Arts, and several high-class homes.

More landfill projects followed until three peninsulas and an island merged to form a single landmass divided by waterways. It is one of the greatest engineering feats in American history.

On the negative side, the influx of people polluted the beaches and sea. Boston Harbor became one of the most polluted harbors in the United States. Cleanup efforts are now underway through the Boston Harbor Project.

Another place where people have had a major effect on a peninsula is in Florida. Here, the Florida Everglades cover the southern part of the state, from Lake Okeechobee to the Gulf of Mexico. Originally, rainwater would feed the rivers, which fed the lake. From the lake, the water seeped into the swamps and marshes of the Everglades.

Then, drawn by the subtropical climate, people moved to Florida. They wanted to live and grow crops on the wetlands, but they didn't want it to be wet anymore. So, the Army Corps of Engineers diverted the water into canals. This prevented flooding and allowed irrigation. Unfortunately, it wreaked havoc on the wetland ecosystem.

Today, 50 percent of the Everglades has been lost forever to urban development and agriculture. Agricultural and lawn chemicals have polluted the lake and the Everglades. Habitat has disappeared. Animals have vanished. Ninety percent of the Everglade bird population is gone, and Florida panthers have dwindled from 1,500 to about 80.

Scientists believe nothing can be done to restore the Everglades to its natural state. However, Floridians want to preserve what is left of the swamps and marshes. The Comprehensive Everglades Restoration Plan, a thirty-year, $7.8 billion project, is designed to change the remaining 50 percent of the Everglades back to how it was.

The Everglades have been changed drastically by people settling in the area. Today, many Floridians work hard to preserve what is left of the unique swamp and marshlands.

Florida and Boston offer lessons to us. Peninsulas, like other coastal areas, are hubs of trade, commerce, and culture, as well as bastions of natural beauty. Wanting to make a living or relax by the sea, people flock to their shores. But unless people take care of the land and oceans, that beauty disappears and coastal economies falter.

Peninsulas have formed because of many natural processes—sedimentation, erosion, volcanism, rising or falling sea levels, and glaciation. Keeping them beautiful, however, depends on just one force of nature: us.

GLOSSARY

bay Part of an ocean, sea, or lake that cuts into the shoreline and is partially enclosed by land.

bog A wetland comprised of decaying plant matter (peat) that is typically found in cooler climates.

cape A narrow piece of land jutting into a body of water.

continent The seven largest land areas on the earth: Africa, Asia, Australia, Europe, North America, South America, and Antarctica.

continental drift The theory that the continents shift their positions by sliding along the molten rocks in the earth's mantle.

debris avalanche An avalanche of mud and rock, not snow.

delta A fan-shaped deposit of sediment that forms when a river enters an ocean or lake.

deposition The process of material (rocks, sand, soil, etc.) being transported and settling someplace new.

erosion The loosening and dissolving of soil and rock from the earth's surface by water, wind, or ice.

geology The study of the earth.

glacier A large mass of flowing snow or ice that forms when extreme snowfall exceeds the rate of melting.

gulf Part of a large body of water that cuts into the shoreline, is partially enclosed by land, and is typically bigger than a bay.

ice age Any geologic time when glaciers covered much of the earth. The most recent was the Wisconsin glaciation.

island Land completely surrounded by water and smaller than Australia (the smallest continent).

isthmus A narrow strip of land that connects two islands, two large landmasses, or a landmass and a peninsula.

limestone Sedimentary rock usually made up of skeletons and shells of marine animals.

marsh A wetland composed of plants like grass and cattails.

ocean The body of salt water covering about two-thirds of the earth; the planet's four major bodies of salt water are the Atlantic, Pacific, Indian, and Arctic oceans.

peninsula Land almost surrounded by water and connected to a larger landmass, sometimes by an isthmus.

Pleistocene era The geologic time period during which glaciers advanced and retreated and humans evolved into their modern form.

sea A body of salt water partially or fully surrounded by land; a huge freshwater lake is also sometimes called a sea.

sediment Fragmental material originating from eroded rocks or organic matter and deposited by water, ice, or wind.

sedimentary rock Stone formed from preexisting rocks or pieces of once-living organisms.

swamp A wetland containing trees or shrubs.

tectonic plates The enormous sheets of slowly moving rock that make up the earth's crust.

FOR MORE INFORMATION

American Geological Institute
4220 King Street
Alexandria, VA 22302-1502
(703) 379-2480
Web site: http://www.agiweb.org
This is a federation of forty-five geoscientific organizations.

Geological Association of Canada
Department of Earth Sciences
Room ER4063
Alexander Murray Building
Memorial University of Newfoundland
Saint John's, NF A1B 3X5
Canada
(709) 737-7660
Web site: http://gac.ca
The Geological Association of Canada promotes the study of geoscience in Canada.

Geological Society of America
P.O. Box 9140
Boulder, CO 80301-9140
(303) 357-1000
Web site: http://www.geosociety.org
This is a society for earth scientists, whether teachers, students, or geologists.

The Nature Conservancy
4245 North Fairfax Drive, Suite 100
Arlington, VA 22203-1606
(703) 841-5300
Web site: http://www.nature.org
This charity protects miles of land and rivers, and operates more than one hundred marine conservation projects globally.

U.S. Geological Survey
12201 Sunrise Valley Drive
Reston, VA 20192
(703) 648-4000
Web site: http://www.usgs.gov
This agency studies the earth and provides reliable information about its landscape, natural resources, and hazards.

Web Sites

Due to the changing nature of Internet links, Rosen Publishing has developed an online list of Web sites related to the subject of this book. This site is updated regularly. Please use this link to access the list:

http://www.rosenlinks.com/lan/penin

FOR FURTHER READING

Anderson, Alan, Gwen Diehn, Terry Krautwurst, Joe Rhatigan, and Heather Smith. *Science Smart*. New York, NY: Main Street, 2003.

Bjornerud, Marcia. *Reading the Rocks: The Autobiography of the Earth*. New York, NY: Basic Books, 2006.

Luhr, James, ed. *Smithsonian Earth*. London, England: DK, 2003.

Pfeffer, Susan Beth. *The Dead and the Gone*. New York, NY: Harcourt Children's Books, 2008.

Pfeffer, Susan Beth. *Life as We Knew It*. New York, NY: Harcourt Children's Books, 2006.

Thompson, David. *Processes That Shape the Earth*. New York, NY: Chelsea House Publications, 2007.

Vogt, Gregory. *Earth's Core and Mantle: Heavy Metal, Moving Rock*. Minneapolis, MN: Twenty-First Century Books, 2007.

Vogt, Gregory. *The Lithosphere: Earth's Crust*. Minneapolis, MN: Twenty-First Century Books, 2007.

BIBLIOGRAPHY

Ciesielski, Paul. "The Day the Mediterranean Dried Up." Retrieved October 14, 2008 (http://www.clas.ufl.edu/users/pciesiel/gly3150/med_story.html).

Comprehensive Everglades Restoration Plan. "FAQs: What You Should Know About the Comprehensive Everglades Restoration Plan (CERP)." Retrieved October 14, 2008 (http://www.evergladesplan.org/facts_info/faqs_cerp.aspx).

Devoli, G., S. P. Schilling, and J. W. Vallance. "Report: Lahar Hazards at Mombacho Volcano, Nicaragua." USGS, 2001. Retrieved August 25, 2008 (http://vulcan.wr.usgs.gov/Volcanoes/Nicaragua/Publications/OFR01-455/introduction.html).

Everhart, Mike. *Oceans of Kansas Paleontology*. Retrieved October 28, 2008 (http://www.oceansofkansas.com).

Florida Department of Environmental Protection. "Florida's Rocks and Minerals." Retrieved October 28, 2008 (http://www.dep.state.fl.us/geology/geologictopics/rocks/florida_rocks.htm).

Funiciello, Renato, Grant Heiken, and Donatella de Rita. *The Seven Hills of Rome: A Geological Tour of the Eternal City*. Princeton, NJ: Princeton University Press, 2005.

Guardian. "North Korea Recent History." Retrieved October 14, 2008 (http://www.guardian.co.uk/flash/0,,855083,00.html).

Hall, Stephen. "Last of the Neanderthals." *National Geographic*, October 2008.

Handwerk, Brian. "New Underwater Finds Raise Questions About Flood Myths." National Geographic News, May 28, 2002. Retrieved August 25, 2008 (http://news.nationalgeographic.com/news/2002/05/0528_020528_sunkencities.html).

Howe, Jeffrey. "Boston: History of Landfills." Retrieved August 1, 2008 (http://www.bc.edu/bc_org/avp/cas/fnart/fa267/sequence.html).

Kunzig, Robert. "Against the Current: Pouring Cold Water on a Climate Myth." *U.S. News and World Report*, May 25, 2003. Retrieved August 25, 2008 (http://www.usnews.com/usnews/culture/articles/030602/2gulfstream.htm).

LaCoast.gov. "Mississippi River Delta Basin." Retrieved October 14, 2008 (http://www.lacoast.gov/landchange/basins/mr/).

Lamb, Simon, and David Sington. *Earth Story: The Shaping of Our World*. Princeton, NJ: Princeton University Press, 1998.

Lovgren, Stefan. "Warming to Cause Catastrophic Rise in Sea Level?" National Geographic News, April 26, 2004. Retrieved August 25, 2008 (http://news.nationalgeographic.com/news/2004/04/0420_040420_earthday_2.html).

Luhr, James, ed. *Smithsonian Earth*. London, England: DK, 2003.

National Park Service. "Geology Field Notes: Cape Cod National Seashore, Massachusetts." U.S. Department of the Interior. Retrieved August 1, 2008 (http://www.nature.nps.gov/geology/parks/caco/index.cfm).

Pyne, Stephen. *Vestal Fire: An Environmental History, Told Through Fire, of Europe and Europe's Encounter with the World*. Seattle, WA: University of Washington Press, 1997.

Samford University. "Wetland Basics." Retrieved August 12, 2008 (http://www.samford.edu/schools/artsci/biology/wetlands/basics/types.html).

Sargent, Bill. "Surf Uncovers Tracks Laid in 1700s." *Boston Globe*, December 14, 2004. Retrieved August 24, 2008 (http://www.boston.com/news/globe/health_science/articles/2004/12/14/surf_uncovers_tracks_laid_in_the_1700s/).

Science@NASA. "Continents in Collision: Pangaea Ultima." October 6, 2000. Retrieved October 14, 2008 (http://science.nasa.gov/headlines/y2000/ast06oct_1.htm).

Science Daily. "Isthmus of Panama Formed as Result of Plate Tectonics, Study Finds." July 31, 2008. Retrieved October 3, 2008 (http://www.sciencedaily.com/releases/2008/07/080729234142.htm).

Sherratt, Andrew. "Portages: A Simple but Powerful Idea in Economic History." *ArchAtlas Journal*. Retrieved August 1, 2008 (http://www.archatlas.org/Portages/Portages.php).

Smithsonian Museum of Natural History. "Types and Processes Gallery: Volcanic Landslides." Global Volcanism Program. Retrieved August 29, 2008 (http://www.volcano.si.edu/world/tpgallery.cfm?category=Volcanic%20Landslides).

Thomas, R. A. "Formation and Evolution of the Delta." Retrieved October 14, 2008 (http://www.loyno.edu/lucec/mrddocs/11.doc).

U.S. Geological Survey. "Frequently Asked Questions: What Are Sedimentary Rocks?" Retrieved August 3, 2008 (http://www.usgs.gov/faq/list_faq_by_category/get_answer.asp?id=503).

VanCleave, Janice. *Earth Science for Every Kid: 101 Easy Experiments That Really Work*. New York, NY: John Wiley & Sons, 1991.

Zeiller, Warren. *A Prehistory of South Florida*. Jefferson, NC: McFarland & Company, 2005.

INDEX

About the Author

Bridget Heos is a children's writer living in Kansas City, Missouri. Her favorite things to write about are science, early history, and anything funny. She has a husband and three sons. Together they have traveled to the Cape Cod Peninsula and the former Shawmut Peninsula (now known as Boston).

Photo Credits

Cover © www.istockphoto.com/Joe Gough; pp. 4–5, 21 Visible Earth/ NASA; p. 7 © AP Images; p. 10 © Bettmann/Corbis; pp. 12–13 Steffen Thalemann/The Image Bank/Getty Images; pp. 14, 35, 50 (bottom) Library of Congress Geography & Map Division; pp. 16–17 NOAA; p. 20 Jan Cobb Photography Ltd./Photographer's Choice/Getty Images; pp. 24–25 © Mike Nelson/epa/Corbis; p. 27 Dennis Flaherty/The Image Bank/Getty Images; p. 29 © Christophe Boisvieux/Corbis; pp. 32–33 © Tom Bean/ Corbis; p. 38 © Eberhard Streichan/zefa/Corbis; pp. 42–43 D'ARCO EDITORI/De Agostini Picture Library/Getty Images; pp. 46–47 © David Littschwager/National Geographic; p. 50 (top) http://www.lib.utexas.edu/ maps/national_parks/boston_nhp99.pdf; p. 53 © Charles Philip Cangialosi/Surf/Corbis.

Designer: Les Kanturek; Editor: Bethany Bryan
Photo Researcher: Amy Feinberg